Naama Ferreira
Clécio D.Dias Silva

Educational Practices in Science and Biology Teaching

Naama Ferreira
Clécio D.Dias Silva

Educational Practices in Science and Biology Teaching

Teaching proposals for basic education

ScienciaScripts

Imprint

Cover image: www.ingimage.com

This book is a translation from the original published under ISBN 978-3-330-99796-7.

Publisher:
Sciencia Scripts
is a trademark of
Dodo Books Indian Ocean Ltd. and OmniScriptum S.R.L publishing group

120 High Road, East Finchley, London, N2 9ED, United Kingdom
Str. Armeneasca 28/1, office 1, Chisinau MD-2012, Republic of Moldova, Europe
Managing Directors: Ieva Konstantinova, Victoria Ursu
info@omniscriptum.com

Printed at: see last page
ISBN: 978-620-8-61974-9

ABOUT THE AUTHORS:

CLÉCIO DANILO DIAS DA SILVA

She has a degree in Biological Sciences from the FACEX University Centre - UNIFACEX (2015). She specialises in Teaching Natural Sciences and Mathematics at the Federal Institute of Education, Science and Technology of Rio Grande do Norte - IFRN (2017). He is studying for a Master's degree in Teaching Natural Sciences and Maths at the Federal University of Rio Grande do Norte (PPGECNM/ UFRN). He is currently working at the UNIFACEX University Centre as a collaborating researcher on Scientific Initiation Research projects (PROIC).

CARMEM MARIA DA ROCHA FERNANDES

She has a degree in Biological Sciences from the FACEX University Centre - UNIFACEX (2014). She specialised in Teaching Natural Sciences and Mathematics at the Federal Institute of Education, Science and Technology of Rio Grande do Norte - IFRN (2016). She is currently working at the FACEX University Centre as a collaborating researcher on Scientific Initiation Research projects (PROIC).

DANIELE BEZERRA DOS SANTOS

He has a degree in Biological Sciences from the Facex University Centre - UNIFACEX (2005). She has a Master's degree in Aquatic Bioecology (2007) and a PhD in Psychobiology (2013), both from the Federal University of Rio Grande do Norte (UFRN). She is currently coordinator of the undergraduate course in Biological Sciences and the postgraduate course in Microbiology and Parasitology at the Facex University Centre, as well as acting as institutional coordinator of the Institutional Teaching Initiation Scholarship Programme (PIBID/CAPES/UNIFACEX).

IVANEIDE ALVES SOARES DA COSTA

He has a degree in Biological Sciences from the Federal University of Rio Grande do Norte - UFRN (1986). She has a specialisation in Aquaculture and Aquatic Bioecology (1996), a

Masters in Aquatic Bioecology (1999), both from UFRN, and a PhD in Ecology and Natural Resources from the Federal University of São Carlos - UFSCAR (2003). She is currently an adjunct professor in the Department of Microbiology and Parasitology at UFRN and coordinator of the PIBID/UFRN Biology sub-area. She is also involved in the postgraduate programmes PRODEMA (Environmental Development Programme) and PPGECNM (Postgraduate Programme in the Teaching of Natural Sciences and Mathematics), also at UFRN.

LÚCIA MARIA DE ALMEIDA

She has a degree in Biological Sciences (1991), a Bachelor's degree in Biological Sciences (1992), and a degree in Artistic Education - Plastic Arts (2001), both from the Federal University of Rio Grande do Norte (UFRN). She has a Masters in Botany from the Federal Rural University of Pernambuco - UFRP (1995) and a PhD in Psychobiology from the Federal University of Rio Grande do Norte - UFRN (2008). She currently teaches undergraduate courses in Biological Sciences, Nutrition and Nursing, and postgraduate courses in Microbiology and Parasitology at the Facex University Centre (UNIFACEX).

NAAMA PEGADO FERREIRA

She has a degree in Biological Sciences from the Federal University of Rio Grande do Norte - UFRN (2012) and a bachelor's degree in Tourism from the University of Potiguar - UnP (2010). She specialised in Teaching Natural Sciences and Mathematics at the Federal Institute of Education, Science and Technology of Rio Grande do Norte - IFRN (2017). She is currently studying for a Master's degree in Teaching Natural Sciences and Maths at the Federal University of Rio Grande do Norte (PPGECNM/UFRN) and is a permanent teacher in the State of Rio Grande do Norte (SEEC/RN).

SUMMARY

PRESENTATION

It is with great satisfaction that we present in this book part of our academic and professional experiences in the science and biology classroom.

This book has been organised to be easy to read, serving as a basis for undergraduates, postgraduates and educators in the field who aim to develop quality teaching. All the dynamics, games and didactic sequences described and used can be easily replicated and adapted to different educational contexts, helping you in the process of planning lessons and serving as a theoretical basis for eventual teaching. This material includes work developed in the field of science and biology teaching:

In the initial chapter *"Concept* maps *and the learning of invertebrate taxa", the* authors propose the use of concept maps as a potentially significant didactic resource for working on morphophysiological, ecological and taxonomic aspects of invertebrate groups.

The second chapter, *"Setting up a school herbarium as a teaching resource for biology"*, presents a Didactic Sequence for working with the various contents of botany (anatomy, physiology and diversity) linked to the students' local contradictions, using the construction of a didactic herbarium in the school environment.

The third chapter *"Alternative* methodologies *in the teaching and learning process of Zoology"* describes and analyses the use of various didactic methodologies (**role-plays**, games, modelling, schematic drawings, dynamic readings, the use of parodies, educational software) for learning about animals, both vertebrates and invertebrates, and their relevance to the ecosystem.

In the fourth chapter, *"The use of comics in the prevention of dengue, zika and chikungunya"*, the authors explore the use of the textual genre of comics to address preventive aspects of endemic diseases caused by mosquitoes of the *Aedes* genus.

The fifth chapter *"Playful* activity *for teaching photosynthesis in secondary school"* presents a didactic game designed to facilitate the learning of photosynthesis, an important biological process for living beings, in a collaborative and participatory way.

In the sixth chapter, *"Didactic models in the teaching of cell biology: a proposal to highlight and overcome alternative conceptions"*, the authors present a didactic and low-cost way of facilitating the learning of cytology content (cell organelles, differences between cells) in order to identify and remedy students' possible alternative conceptions on the subject.

The chapter "*Play to encourage reading and assess learning in science teaching*" presents a teaching dynamic to revise science content in a participatory and collaborative way, with aspects to encourage students to read.

In the last chapter, "The *use of pedagogical workshops as a didactic resource in sexuality education*", the authors present a didactic sequence aimed at demystifying questions about sexuality, in an open way, with adolescents, so that they could reflect on teenage pregnancy, STDs, among other topics.

We hope that this book will serve as an incentive to improve your future classes, making them ever more dynamic, participatory and creative. May you feel encouraged not only to learn, but to share these and other didactic experiences with us in favour of quality education for our country, training scientifically literate students.

The organisers

CHAPTER 1

CONCEPT MAPS AND LEARNING ABOUT INVERTEBRATE TAXA

Clécio Danilo Dias da Silva[1]

Carmem Maria da Rocha Fernandes[2]

Daniele Bezerra dos Santos[3]

Lúcia Maria de Almeida[4]

INTRODUCTION

In school curricula, Zoology (the study of animals) is currently linked to the disciplines of Science in Primary School and Biology in Secondary School, and it is through Zoology that the history of animals, in all its aspects, has been taught (SILVA et al., 2016). With regard to the approach to Zoology in the learning environment, Pereira (2012) points out some problems that interfere with the quality of the teaching and learning process of this subject, such as: the prevalence of creationist ideas and religious conceptions that mix with scientific knowledge; the poor initial training of teachers that does not provide an adequate basis for working on the subject; lack of practical classes on Zoology subjects, appropriate laboratories, lack of teaching materials and lack of knowledge of Zoology teaching techniques.

In view of this, Bastos Junior (2014) states that the use of playful activities such as role-plays, games, modelling, schematic drawings, dynamic readings, the use of parodies, educational software and the creation of concept maps can help to minimise these problems, providing meaningful learning of the various contents of Zoology.

According to Trindade (2011), a necessary condition for meaningful learning to occur

1 Specialist in Science Teaching (IFRN) and Master's student in Natural Science Teaching (UFRN).
2 Specialist in Teaching Natural Sciences and Mathematics (IFRN).
3 Master in Bioecology (UFRN) and PhD in Psychobiology (UFRN).
4 Master in Botany (UFRPE) and PhD in Psychobiology (UFRN).

is for the material to be learnt to be potentially significant, i.e. relateable (or incorporable) to the learner's cognitive structure. And, referring to concept maps (CM), Moreira (2010) defines them as diagrams indicating relationships between concepts, or between words that we use to represent concepts. Complementing this thought, Gomes et al. (2010) points out that CMs are dynamic and flexible instructional tools, used both in the analysis and organisation of content, which become instruments that favour the association and interrelationship between old and new concepts, as advocated by David Ausubel's Significant Learning Theory (SLT) (AUSUBEL, 2003).

With this in mind, the aim of this study was to analyse the use of Concept Maps in the teaching and learning process of the knowledge covered in invertebrate zoology in basic education.

METHODOLOGY

The research took place at the Soldado Luiz Gonzaga State School, located in the urban area of Natal, in the state of Rio Grande do Norte, with 22 students from the 2nd year of secondary school. Firstly, a process of familiarisation with concept maps was carried out using readings and discussions of didactic material containing norms, steps and suggestions for constructing concept maps.

In the second stage, there were lectures and dialogues with the aid of multimedia resources, enabling students to deepen and systematise their knowledge of two invertebrate taxa (Annelida and Mollusca), covering points relating to the morphology, physiology, ecology and phylogeny of these groups. At the end of each lecture, the contents were summarised using previously prepared MCs, with the aim of giving the students an introduction to the technique.

In the third stage, the students were asked to draw up a concept map for each taxon studied. To help the students, they were given didactic texts previously prepared with the help of appropriate literature (AMABIS; MATHO, 2010, LOPES; ROSSO, 2014), providing information on the two groups studied.

The concept maps produced by the students were analysed according to the assessment criteria proposed by Trindade (2011) (Table 1). This is an assessment guideline that considers qualitative and quantitative aspects by establishing scoring categories, in some of which significant changes to the structure of the maps are sought. The scoring range for each category varied from 0 to 1 point, distributed as follows: 1 (correct), 0.5 (partially correct) and 0 (error). Taking into account the existence of 10 categories, the total number of points allowed for each map was 10. We used the 50% standard as the satisfactory average (SA), i.e. half of the total allowed, 5.0 points; and unsatisfactory average (MI) scores below the standard, as proposed by Lourenço (2008).

CATEGORIES	DESCRIPTION OF THE ASSESSMENT CRITERIA
1st basic concepts	Does the map contain at least 50 per cent of the basic concepts from the list provided and/or the reference map?
2nd new concepts	Are there any new concepts relevant to the subject in question?
3rd links between concepts	Are all the concepts connected by well-made lines?
4th linking words (connectives)	Do most linking words/phrases make logical sense with the concept they link to?
5th examples	Does the map show appropriate examples for the subject in question?
6° clarity and aesthetics of the map.	Is the map legible and easy to read? Is the map clear to the reader? Is the map legible, without scratches or smudges? Is the handwriting legible? Do all the concepts appear in boxes? Is the spelling correct?
7th proposition (linking word concept)	Does the map have at least 50 per cent of the number of valid propositions of the reference map? Do the propositions have logical meaning from a semantic and scientific point of view? Are the connections in line with what is scientifically accepted?
8th prioritisation	Is there a successive ordering of the concepts? Is there a good hierarchisation of the concepts, represented by at least 03 hierarchical levels?
9th progressive differentiation	Is it possible to distinguish between more inclusive and subordinate concepts? Can you clearly identify the most general and the most specific concepts? Is there a progressive conceptual differentiation that shows the degree of subordination between concepts?
10th integrative reconciliation	Is there a recombination, i.e. a rearrangement of concepts? Are there cross-cutting relationships between concepts belonging to different parts of the map?

Table 1: Categories for analysing concept maps proposed by Trindade (2011).

In general, the data obtained was grouped and categorised in tables in the Microsoft Excel 2010 application, in order to draw up tables and construct the results and discussions.

RESULTS AND DISCUSSIONS

According to Ontoria Pena et al. (2005), Concept Maps enable meaningful learning to the extent that they are drawn up by the students and used as a tool for appropriating knowledge. In view of this, the materials produced by the students allowed us to see that CMs can be considered a potentially significant tool for learning zoology content, since, when dealing with the Mollusca group, out of a total of 22 CMs, 17 (77%) obtained a Satisfactory average, and only 5 (23%) obtained an Unsatisfactory average. As for the

Annelida group, there was a slight increase in the number of Concept Maps with averages considered satisfactory, since 20 (91%) of the CMs obtained a satisfactory average, and only 2 (9%) obtained an average below expectations (Table 2). These results are in line with those proposed by Lourenço (2008), since the majority of students obtained a MS in both CMs (77% and 91%).

MOLLUSCA PHYLUM											
STUDENT	**AI**	**A2**	**A3**	**A4**	**A5**	**A6**	**A7**	**A8**	**A9**	**AIO**	**A11**
NOTE	7,0	3,5	**5,5**	5,0	5,5	6,0	**7,5**	8,0	4,5	5,0	6,0
STUDENT	**A12**	**A13**	**A14**	**A15**	**A16**	**A17**	**A18**	**A19**	**A20**	**A21**	**A22**
NOTE	8,5	4,5	**6,5**	4,5	5,0	8,5	**5,5**	7,0	6,5	5,5	4,0
PHYLUM ANNELIDA											
STUDENT	**AI**	**A2**	**A3**	**A4**	**A5**	**A6**	**A7**	**A8**	**A9**	**AIO**	**A11**
NOTE	8,0	6,5	**7,5**	7,0	6,0	8,0	**6,5**	7,0	3,0	4,0	7,5
STUDENT	**A12**	**A13**	**A14**	**A15**	**A16**	**A17**	**A18**	**A19**	**A20**	**A21**	**A22**
NOTE	9,0	6,0	**9,5**	8,0	6,0	9,0	**5,5**	8,5	7,0	5,5	6,0

Table 1: Grades given to the concept maps drawn up by the students.

In relation to the Evaluation Categories (Figure 1), we found that in both the MCs dealing with the Mollusca group and the Annelida group, the students showed ease respectively in the categories of: Basic Concepts (79% and 89%), Hierarchisation (63% and 79%), Linking Words (66% and 72%), Progressive Differentiation (66% and 58%), Linking Concepts (59% and 61%), New Concepts (62% and 60%) and Clarity and Aesthetics of the CMs (52% and 55%).

According to Trindade (2011), these categories include the easiest points to construct maps, which would justify the ease with which the students constructed their CMs. However, we hope that the students' good performance in these categories should not be associated with ease, but rather with the appropriation and meaningful learning of the content that was worked on, deepened and discussed in the learning environment.

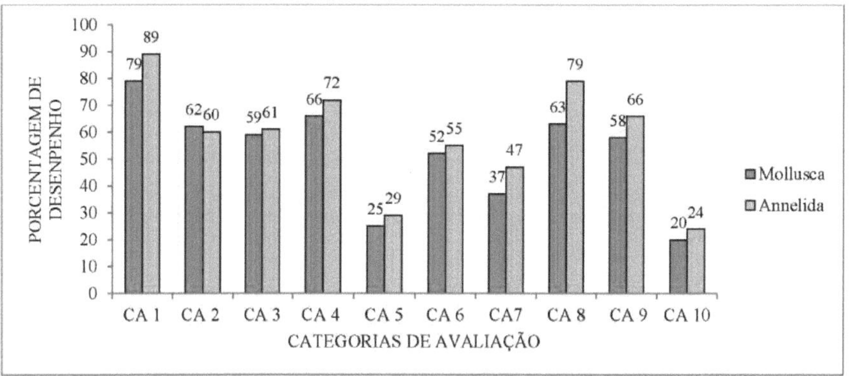

Figure 1: Frequency of performance identification, in percentages, of the categories.
Key: CA 1 (basic concepts), CA 2 (new concepts), CA 3 (link between concepts), CA 4 (linking words), CA 5 (examples), CA 6 (clarity and aesthetics of the map), CA 7 (propositions), CA 8 (hierarchisation), CA 9 (progressive differentiation), CA 10 (integrative reconciliation).

On the other hand, the categories related to Propositions (37% and 47%) Examples (25% and 29%) and Integrative Reconciliation (20% and 24%) were the ones that students had the most difficulty developing in their CMs. The students' embarrassment with the examples and integrative reconciliation when drawing up Concept Maps may be related to the fact that the learners consult little instructional material, or look for sources of information other than that provided by the teachers, limiting themselves to the knowledge acquired in the classroom. With regard to the propositions, we found that the students' greatest difficulties were in making logical sense of the concepts, especially those relating to the ecological and phylogenetic aspects of the groups. Generally speaking, one way of alleviating these limitations is by revisiting the topics raised during lessons, with the aim of assimilating the new information, as Ausubel (2003) advocates.

FINAL CONSIDERATIONS

The positive results obtained through this research confirm the efficiency of using Concept Maps to address Invertebrate taxa within zoology content, since the students obtained significantly satisfactory averages in the CMs produced, addressing the Mollusca and Annelida groups. We infer that CMs can be considered a potentially significant didactic tool for learning zoology content and can be applied to science and biology subjects in basic education.

REFERENCES

AMABIS, J. M.; MARTHO, G.R. **Biologia**. São Paulo: Moderna, 2010.

AUSUBEL, D. P. **Acquisition and Retention of Knowledge:** a cognitive perspective. Lisbon: Plátano, 2003.

BASTOS JÚNIOR, P. S. **Methodologies and strategies used to teach zoology.** 2013. 24 f. Monograph (Degree in Natural Sciences), University of Brasília, Planaltina - DF, 2013.

GOMES, A. P. et al. Teaching Science: Dialoguing with David Ausubel. **Revista Ciências & Ideias**, v.1, n.1, p. 23-31, 2010.

LOPES, S.; ROSSO, S. **Biologia**. 3ª ed, São Paulo: Saraiva, 2014.

LOURENÇO, A. B. **Analysis of concept maps drawn up by students in the 8th grade of primary school after lessons based on the theory of meaningful learning**: clay as a subject of study. 2008. 115 f. Dissertation (Master's Degree), Federal University of São Carlos, São Carlos - 2008.

MOREIRA, M. A. Concept maps as tools to promote progressive conceptual differentiation and integrative reconciliation. **Ciência e Cultura**, v.32, n.4, p.474-479, 2010.

ONTORIA PENA, A. et al. **Conceptual Maps**. São Paulo: Loyola, 2005.

PEREIRA, N. B. **Perspectives for teaching zoology and possible directions for a different practice from the traditional one**. 2012. 43 f. Monograph (Biological Sciences), Mackenzie Presbyterian University - SP, 2012.

SILVA, C. D. D et al. Zoological caravan: contributions to science and biology teaching. In: National Education Congress, 3, 2016. **Proceedings...** Natal: Realise Eventos e Editora, 2016.

TRINDADE, J. D. **Teaching and meaningful learning of the chemical bond concept through concept maps**. 2011. 230 F. Dissertation (Professional Master's Degree in Chemistry), Federal University of São Carlos, São Carlos - SP, 2011.

CHAPTER 2

ASSEMBLING A SCHOOL HERBARIUM AS A DIDACTIC RESOURCE FOR TEACHING BIOLOGY

Clécio Danilo Dias da Silva[5]

Carmem Maria da Rocha Fernandes[6]

Naama Pegado Ferreira[7]

Daniele Bezerra dos Santos [8]

Lúcia Maria de Almeida[9]

INTRODUCTION

In the teaching of Botany, especially in the content of Plant Systematics and Taxonomy, it is common to use a very wide range of terms, either to name and describe the different structures of plants and/or to indicate the different species and groups.

This model does not seek to build knowledge or even arouse students' interest in the content covered in the learning environment, since this teaching is centred on lectures based solely on "science textbooks" (KRASILCHIK, 2011).

In view of this, there is a clear need to develop actions that enable students to understand botany in a critical and meaningful way, such as recognising plants in the school's surroundings, or even relating plants in the environment from a holistic viewpoint, i.e. addressing plant-animal and plant-man relationships, among others. For Bitencout (2013), a working method that can be used in the area of botany can be a school herbarium.

According to Araújo and Silva (2015), it is a valuable strategy for students to recognise plants in a particular place or region, and especially for developing various biology concepts

[5] Specialist in Science Teaching (IFRN) and Master's student in Natural Science Teaching (UFRN).
[6] Specialist in Teaching Natural Sciences and Mathematics (IFRN).
[7] Specialist in Science Teaching (IFRN) and Master's student in Natural Science Teaching (UFRN).
[8] Master in Bioecology (UFRN) and PhD in Psychobiology (UFRN).
[9] Master in Botany (UFRPE) and PhD in Psychobiology (UFRN).

by manipulating plants and their structures in a way that makes learning more engaging and exciting for students.

Silva, Almeida and Romeiro (2016) state that the school herbarium is a resource that allows the teacher to make all the necessary adaptations to meet their particular or local needs, making it an excellent teaching tool, as it contributes to knowledge of collection techniques, mounting of exsiccates, systematics, morphological and taxonomic studies, as well as the development of interactive keys for the identification of botanical groups (BURITY; MIGUEL; JASCONE, 2011).

METHODOLOGY

This work is characterised by a qualitative approach using action research, the aim of which is to provide information to guide decision-making and improve practice, where in the course of the work the research itself becomes action, contributing to the articulation between theory and practice (ESTEBAN, 2010).

The data was collected during the interventions and socialisations in the focus groups (selected classes) where the research was carried out. The activities took place from February to June 2015 at the José Fernandes Machado State School, located in Natal, Rio Grande do Norte. A total of 45 1st year students and 51 2nd year students took part in the educational activities, totalling 96 participants.

In order to develop the school herbarium, a didactic sequence was constructed covering five stages (Figure 1). Initially, a survey was carried out on their knowledge of botany by asking them what they knew about the plants in the school, using the following questions: "*Have* you *ever stopped to look at the plants in the school's flowerbeds and can you identify and name them?"*.

After the survey, lectures and dialogues were held, as well as discussion groups, where content such as the concept of the Plantae kingdom, its main characteristics, and the division of groups (bryophytes, pteridophytes, gymnosperms and angiosperms) were discussed. At this stage, a greater focus was placed on the morphology and taxonomy of angiosperms, as these were of fundamental importance for the following stages. Later, in order to better understand the characteristics of the plants in this group, practical lessons were held in the school's science laboratory, where the students were able to observe and recognise the basic structures of flowers (petals, sepals, stamens, pistil, receptacle, etc.) and leaves (petiole, stipules, limbus, sheath, veins, etc.).

Figure 1: Steps taken to develop the School Herbarium.

After these steps, the students were given a lecture, presenting the proposal to develop the Didactic Herbarium. At this point, the students were shown what a herbarium is, how it is built, its scientific functions and its potential within a school environment. The basic procedures for collecting, identifying and herborising the plants that would be collected were also discussed with the students.

The classes were then divided into pairs to collect and identify the plants. Initially, the pairs were taken on a field trip around the school, where they visually analysed the species present on site. Afterwards, the students collected the specimens they had observed. Each pair was responsible for collecting at least one plant. During this stage, the students were monitored and helped by the teachers involved in the activities to remove parts of the plant (small branches, leaves and flowers) so that it could be collected as completely as possible, making it easier to recognise and identify the plants in the future.

During this process, the students wrote down the data of the individuals observed in the field (vegetative habits, shapes, colouring, etc.).

Material was collected from 47 plants to build the local school herbarium. After this, together with the teacher and scholarship monitors, the students were taken to the laboratory, and with the help of magnifying glasses and identification keys from Souza and Lorenzi (2009), they identified the plants down to the smallest possible taxon. The information was entered onto an "identification sheet". Teachers, monitors and students identified a total of 25 species, distributed among 15 families, culminating in Table 1.

Following routine procedures for preparing exsiccates, branches from the different plants found around the school were soaked in 70% alcohol to prevent the loss of structures, as the solution helps to fix the plant parts. Once this was done, the collected material was protected between sheets of folded newspaper, which were placed between cardboard, forming layers, closing the pile in the form of a "sandwich" (cardboard + newspaper + plant + newspaper + cardboard). These were placed inside a wooden press measuring 42 x 30 cm. Twine was used to compress the pile formed by overlapping the plants, after which they were taken to dry, which lasted approximately 2 to 3 days.

FAMILY	SCIENTIFIC NAME	COMMON NAME
AMARANTHACEAE	*Altemantaeratenella* Colla	Stone breaker
ANACARDIACEAE	*Anacendium occidentale* L. *Mansifera indica* L.	Cashew tree Hose
APOCYNACEAE	*('aiharaoShus roseus* *Nerium oleander* L. *Plumeria pudica* Jackfruit.	Good evening Spiraliser Bridal bouquet
ARECACEAE	*Cnuos nucifera* L. *Livistona chinensis* (Jacq.) R. Br. Ex Mart. *Dypsis lutescens* *Rovstonea oleracea* (Jaca.) 0. F. Cook.	Coconut Fan Palm Areca palm Imperial Palm
ASPARAGACEAE	*aansenieria cylindrica* Prain *Sansevieria trifasciata* Prain.	St George's Spear St George's sword
BIGNONIACEAE	*Tabebuia aarea* (Silva Manso)	Craibeira
CHRYSOBALANACEAE	*Tscanie toaentesa* (Benth.) Fritsch	Oitizeiro
COMBRETACEAE	*aerminalia catappa* L.	Chestnuts
FABACEAE	*Caesalpinia echinata* Lam. *Cassia fistula* L. *Prosopis iuliflora* (SW) D.C.	Brazilwood Acacia Algaroba
HELICONIACEAE	*POeliconia biaai*	Heliconia
MALVACEAE	*aauatra aquatica* Aubl. *Talipariti tiliaceum* (L.) Frvxell	Mungubeira Beach Cotton
MELIACEAE	*azndirachta indica* A. Juss.	Nim
MYRTACEAE	*Svzveium malacre&se* (L.) Merr. & L. M.	Jambeiro

	Perrv.	
RUBIACEAE	*cxora coccinea* L.	Ixora
TURNERACEAE	*Turnera ulmifelia* L.	Chanana

Table 1: Species collected and present in the school herbarium.

During the drying period, there were a few changes of newspaper used to absorb the water, so that the humidity would not damage the material. At this point, the students researched information about the pressed species in order to better compose the labels that would later be placed on the exsiccates.

After the material had dried, it was glued onto cardboard in proportion to its size. In the bottom corner of the card, the plant data was placed and stored in a mothballed box in the science laboratory of the local school.

RESULTS AND DISCUSSIONS

We realised during the knowledge survey stage that the students have a basic knowledge of the subject, which can be seen in the plants mentioned by the students during the problem. Various species were mentioned that are present in the school environment, such as: cashew, coconut, palm, Xanana, jambeiro, castanhola, mango, beach parsley, etc.

According to Emerich (2010), it is through the initial survey that the teacher is able to identify the basic knowledge that the students have, so that they can adapt their teaching planning according to the learning needs of the class, and also allows them to assess and monitor the progress of the students during their teaching practices .

During the course and organisation of the round table discussions, we found that the students' interest was piqued, intensifying their learning about the subject. They participated effectively, expressing opinions and asking questions throughout the discussions and showing great interest in answering the questions posed during the activities.

Activities involving dialogues, discussions and conversation circles facilitate students' cognitive development, as well as contributing to the learning of science and biology content, enabling the construction of scientific concepts aimed at developing competences that help students deal with information, understand it, rework it, refute it, and thus understand the world and act autonomously in it.

We realised that the students, moments before going out into the field to collect the botanical material, were a little repulsed and averse to the activities, due to the fact that they "had to go *into the middle of the bush"* to collect plants, and especially to having to use various "*complex* names" (the scientific names of the plants). However, this scenario changed as the activity went on, as coming into contact with the plants in the field was fundamental in awakening their interest in botany-related content.

In this sense, Araújo and Silva (2015) state that methodological proposals involving collection and identification with students are essential to help build knowledge, as well as giving them the opportunity to get closer to and interact with the scientific world. In this way, the author also states that when green areas are used as a teaching resource during lessons outside the learning environment, they become a differentiated working tool for the teacher and, at the same time, contribute to the learning process, as the construction of knowledge through activities like this becomes much more meaningful to the students.

During all the stages (collection, pressing and identification), we saw the students' motivation in relation to the activities carried out, culminating in the creation of the school herbarium. According to Araújo and Silva (2015), when working with botany in basic education, one of the activities that stimulate students can be the use of collection, identification, assembly of exsiccates, and consequently the assembly of a school herbarium. In this way, Burity, Miguel and Jascone (2011), Braz and Lemos (2014) developed a didactic herbarium in public schools, and proved a significant improvement in the fixation of this content in the classes where these activities were worked on.

At the end of the activities, we realised the importance of actions such as setting up a school herbarium, as the students began to recognise the various species of flora in the school environment. We also found that the same students who had difficulties with botany-related content began to understand it in a meaningful way, using various technical terms for structures and even scientific names. Authors such as Ota (2012) and Nunes et al. (2015), when developing activities with didactic herbaria in basic education, also found positive results after effective actions, since the students showed significant changes in the words used for plant structures and scientific names of the various plants collected.

FINAL CONSIDERATIONS

The development of activities involving field practice, taxonomic identification and herbarium construction works as an excellent tool for arousing student interest, especially in the teaching of botany, contributing to the construction of student knowledge, adding to the values of meaningful learning, since they arouse curiosity and stimulate the application of these concepts in everyday life. From this perspective, we infer that the use of methodologies such as this is valid and can be associated with the teaching of science and biology in basic education.

REFERENCES

ARAÚJO, J. N.; SILVA, M. F. V. Significant Learning of Botany in Natural Environments. **Revista Areté**, v.15, n.8, p.12-23, 2015.

BITENCOURT, I. M. A. **Botany in High School: Analysis of a Teaching Proposal Based on the CTS Approach**. 2013. Dissertation (Master's Degree); State University of Southwest Bahia, Jequié/BA, 2013.

BRAZ, N. C. S.; LEMOS, J. R. "School herbarium" as a didactic tool for learning about plants in a high school in the city of Parnaíba, Piauí. **Revista Didática Sistêmica**, v.16, n.2, p. 3-14, 2014.

BURITY, C. H. F.; MIGUEL, R. J.; JASCONE, C. E. Creation and application of a didactic herbarium in a state school in the municipality of Duque de Caxias, RJ. **Saúde & Ambiente em Revista,** v.6, n.1, p. 33-45, 2014.

EMERICH, C. M. **Science teaching:** a proposal to adapt knowledge to everyday life - a focus on water. 2010. Master's dissertation. Federal University of Rio Grande do Sul/RS, 2010.

KRASILCHIK, M. **Prática de Ensino de Biologia**. São Paulo: Editora vivaz, 2011.

NUNES, M. J. M. et al. Didactic herbarium as a differentiated tool for learning in a high school. **Revista Diálogos em Educação**, v.24, n.2, p.41-55, 2015.

OTA, M. D. **Herbário escolar:** uma proposta de atividade prática para o ensino de botânica. Monograph (Degree in Biological Sciences), Faculty of Education and Arts, University of Vale do Paraíba, São José dos Campos, SP, 2012.

SILVA, C. D. D.; ALMEIDA, L. M.; ROMEIRO, D. H. L. Learning botany through the three pedagogical moments. In: Encontro Nacional de Pesquisas e Práticas em Educação, 2., Natal. **Proceedings of the National Meeting of Research and Practices in Education**. Rio Grande do Norte, Natal: EDUFRN, 2016.

SOUZA, V. C.; LORENZI, H. **Botânica Sistemática:** Guia ilustrado para identificação das famílias de Fanerógamas nativas e exóticas no Brasil, baseado em APG II. 2. Ed. Nova

Odessa, SP: Instituto Plantarum, 2009.

Odessa, SP: Instituto Plantarum, 2009.

CHAPTER 3

ALTERNATIVE METHODOLOGIES IN THE TEACHING AND LEARNING PROCESS OF ZOOLOGY

Clécio Danilo Dias da Silva[10]

Carmem Maria da Rocha Fernandes[11]

Daniele Bezerra dos Santos[12]

Lúcia Maria de Almeida[13]

INTRODUCTION

According to Hickman, Roberts and Larson (2011) Zoology is the branch of science that seeks to study, classify and understand animal diversity. With regard to school curricula in basic education, this knowledge is covered in the subjects of Science and Biology, in primary and secondary education respectively.

According to Amorim (2002), zoology is often seen as "outdated" in its more morphological approach. However, much of this view is due to the way it is approached in the classroom, as zoology teaching is still made up solely of the presentation of taxonomic groups and the sets of characteristics of individuals, leading to limitations in the contextualisation of zoology.

In addition to the content-based and decontextualised approach commonly used in Zoology content, Pereira (2012) highlights some problems that influence the teaching-learning process of this subject, such as: lack of practical and field classes, appropriate laboratories, lack of teaching materials, lack of knowledge of Zoology teaching techniques, among others.

In order to minimise these problems, Bastos Junior (2014) proposes the use of

[10] Specialist in Science Teaching (IFRN) and Master's student in Natural Science Teaching (UFRN).
[11] Specialist in Teaching Natural Sciences and Mathematics (IFRN).
[12] Master in Bioecology (UFRN) and PhD in Psychobiology (UFRN).
[13] Master in Botany (UFRPE) and PhD in Psychobiology (UFRN).

alternative methodologies, such as role-plays, games, modelling, schematic drawings, dynamic readings, the use of parodies and educational software, to promote meaningful learning of zoology content.

In view of this, the aim of this work was to evaluate the use of alternative methodologies in the teaching and learning process of knowledge relating to the invertebrate and vertebrate groups, emphasising their evolutionary and taxonomic importance, awakening students' critical reflexive sense in relation to biodiversity and the importance of animals within the ecosystem.

METHODOLOGY

DEVELOPMENT OF ACTIVITIES

The activities were carried out between March and June 2014 in a public school in the urban area of Natal - RN, with four primary school classes and three secondary school classes, totalling 170 students.

Initially, the students' previous knowledge of the subject was probed through the "Initial Problematisation" stage. It was developed through the following contextualised and problematising questions: "When you walk through the streets, squares, beaches with family and/or friends, you can observe a great diversity of animals, which are the most common ones you see?", "Could you characterise them? Would you be able to form groupings according to any criteria of similarity between them? Do you know of any groups you've seen in your studies?".

Afterwards, lectures and round table discussions were held on the morphophysiological aspects, habitats, eating habits and ecological relationships of the various phyla and groups of invertebrates and vertebrates. This time was spent using audiovisual resources, videos and interactive models, using alternative materials that are easily accessible and inexpensive and can be found in the students' day-to-day lives.

The students were then directed to the "zoological tents" where specimens of animals were displayed in wet (70% alcohol) and dry (taxidermised) form for each phylum/group of invertebrates (Porifera, Cnidaria, Platyhelminthes, Mollusca, Annelida, and Echinodermata) and vertebrates (Actnopterigii, Sarcopterigii, Amphibia, Reptilia, Aves and Mammalia) discussed above. As they passed by the benches, they passed on some information about each specimen on display, making it easier for the students to understand and retain the knowledge acquired during the lectures and discussions.

After these moments, the students were taken to an open space in the school to take part in the chain dynamics. First of all, some boards were distributed, with one student representing the sun and the other participants symbolising some organism. The student who represented the sun started the activity by throwing the rolling pin to a participant who represented a certain photosynthesising organism. Following this dynamic, the roll was thrown to a herbivore and then to a carnivore. The latter threw the reel to a decomposer, which returned to the sun to restart the cycle. When the reel was passed on to the decomposer, the teacher intervened to explain and clarify any doubts about the influence of that individual's death on the whole web.

We emphasised that in order to recall the zoological groups, the students representing herbivores and carnivores had to name which animal it was and which group it belonged to. In general, the aim of the dynamic was to reinforce the students' knowledge of animal groups and their ecological relationships within the food web.

ANALYSING THE ACTIVITIES CARRIED OUT

At the end of the activities, the students were asked to evaluate the methodology used in the research and its contributions to the process of learning the content covered. To do this, we used the scale proposed by Rensis Likert (LIKERT, 1932) as a collection tool. This material contained 05 statements about the activities carried out (Table 1), and a list of sentences for which the research subjects could express their degree of agreement by ticking: Totally Agree (TC); Partially Agree (CP); Indifferent (IN); Partially Disagree (PD); Totally Disagree (DT). According to Zanella, Seidiel and Lopes (2010), this resource is commonly used in surveys of attitudes, opinions and evaluations, and has made a significant contribution to adding reliability to research in the field of education.

In general, the data obtained was grouped and categorised in tables in the Microsoft Excel 2010 application, in order to draw up graphs and tables and construct the results and discussions.

RESULTS AND DISCUSSIONS

According to Araújo-de-Almeida et al. (2011), the teaching-learning process in zoology is more effective when there is integration between content and methodological strategies, especially if the teaching tools used in the classroom are the result of interactions between students and teachers. In view of this, we believe that student evaluations of the methodologies used in the classroom, providing feedback on these, will enable teachers to improve their pedagogical practices in the school context.

In this sense, in order to find out how effective the alternative methodologies used in this work were, we analysed the students' answers, obtained through the questions in the evaluation tool. These allowed us to check the students' opinion of the activities carried out (Table 1).

For the first statement, 145 students (85%) Totally Agree that "It is *easier and more interesting to learn zoology content using activities that make the lesson more dynamic",* 20 students (12%) Partially Agree with the statement, and only 05 (3%) Totally Disagree. According to Pereira (2012), the use of alternative teaching methodologies attracts attention and arouses students' interest in understanding the various concepts worked on in zoology, enabling more dynamic and meaningful learning of these contents in the classroom.

With regard to the second statement, 110 students (65%) Totally Agree that "The *problematising questions used in the classes contain relations with my daily life, and instigated me to seek knowledge about the biological characteristics of animal groups",* 15 (12%) Partially Agree, 07 (5%) Indifferent, 17 (15%) Totally Disagree and only 05 students (3%) Partially Disagree. According to Gehlen (2009), it is through problematisation at the start of activities that students are challenged to expose their understanding of certain significant situations, doubts, uncertainties, concepts and understandings about the topic being addressed.

Complementing this thought, Emerich (2010) states that the initial survey enables the teacher to identify the students' prior knowledge and thus adjust their teaching planning according to the learning needs of the class, while also making it possible to monitor and evaluate the students' progress during their teaching activities.

AFFIRMATIVE	CT	CP	IN	DT	DP
1. It's easier and more interesting to learn zoology content using activities that make the lesson more dynamic.	145 (85%)	20 (12%)	00 (0%)	05 (3%)	00 (0%)
2. The problematising questions used in the lessons were related to my everyday life and encouraged me to seek knowledge about the biological characteristics of animal groups.	110 (65%)	15 (12%)	07 (5%)	17 (15%)	05 (3%)
3. The lectures and discussions helped me to understand the main characteristics of the taxa covered, such as morphological, physiological, ecological and evolutionary aspects.	90 (53%)	40 (23%)	18 (11%)	05 (3%)	17 (10%)
4. The exhibition of biological material in zoological tents enabled me to better understand the knowledge acquired in the classroom.	160 (94%)	07 (04%)	03 (2%)	00 (0%)	01 (0%)
5. The chaining dynamic enabled me to reinforce the students' knowledge of animal groups and their ecological relationships within the food web.	125 (74%)	19 (11%)	11 (6%)	05 (3%)	10 (6%)

Table 1: Evaluation of the methodology steps. Legend: CT (Totally Agree), CP (Partially Agree), IN (Indifferent), DT (Totally Disagree), and DP (Partially Disagree).

In the third statement, 90 students (53%) Totally Agree that "The lectures and discussions helped me to understand the main characteristics of the taxa covered, such as morphological, *physiological, ecological and evolutionary* aspects", 40 (23%) Partially Agree, 18 (11%) Indifferent, 05 (3%) Totally Disagree and 17 (10%) Partially Disagree. According to Emerich (2010), activities involving dialogues, discussions and conversation circles facilitate students' cognitive development, as well as contributing to the learning of science and biology content, enabling the construction of scientific concepts aimed at developing competences that help students to deal with information, understand it, rework it, refute it, and thus understand the world and act autonomously in it. In this sense, authors such as Bastos Junior (2014) and Silva et al. (2016), when developing activities involving dialogic lessons and discussions on zoology content, found that students' interest was piqued, intensifying learning and systematising their knowledge.

With regard to the fourth statement, 160 students (64%) Totally Agree that "The *exhibition of biological materials in zoological tents enabled me to better understand the knowledge acquired in the classroom."* 7 (4%) Partially Agreed, 3 (2%) Indifferent, and 1 (0%) Partially Disagreed. According to Silva et al. (2016), the exhibition of biological materials is extremely relevant to the process of learning about zoological taxa, arousing students' interest and curiosity, since many of them do not have contact with most of the animals studied in the classroom, such as sponges, molluscs, foxes, penguins, alligators, peacocks, etc; they are only heard and/or seen through textbooks, the media and science communication.

Morais et al. (2015) and Camelo et al. (2015), when developing similar activities in public schools in Natal, RN, found that the use of biological materials is beneficial both for facilitating the understanding of science content and for arousing students' motivation to feel like an integral part of nature, actively contributing to the conservation of local fauna and promoting environmental education. Oporto et al. (2015) also developed activities in schools in Natal - RN and found that these are relevant and stimulate students' curiosity, making zoology content more interesting to students, contributing to better knowledge building in science and biology subjects.

For *THE* fifth statement, 15 students (74%) Totally Agreed that, "The dynamics of chaining enabled me to reinforce the students' knowledge of *animal groups and their ecological relationships within the food web.",* 19 (11%) Partially Agreed, 11 (6%) Indifferent, 05 (3%), and 10 students (6%) Partially Agreed. According to Campos (2010), dynamics are methodological tools that favour students' ability to form concepts, relate ideas, establish logical relationships and develop oral expression. According to Campos (2010), dynamics are methodological tools that favour students' ability to form concepts, relate ideas, establish logical relationships and develop oral expression. For the author, they also enable students to develop initiative, reflective capacity, teamwork skills, assimilation, problem solving and the development of autonomy.

In general, given the variety of didactic experiences involving zoology and the unanimous acceptance of the strategies by the students, we can infer the potential of the alternative methodologies described here for working with zoological taxa in basic education.

CONCLUSION

By carrying out the proposed methodologies, we found that the use of conversation circles, zoological tents and dynamics works as an excellent tool for arousing students' interest, "escaping" from the mechanistic and content-based method commonly used in zoology teaching.

We found that the use/exhibition of biological materials in the school environment is relevant to students' learning, since they allow students to relate these materials to the content seen in class. We also realised how effective they are in the teaching-learning process when compared to the images and diagrams used in theory classes, since they enable students to have "real contact" with species that may be/are present in their daily lives, unlike the simplistic use of images in textbooks.

From this perspective, we infer that the use of methodologies such as this is valid and can be linked to the teaching of zoology, being effective and making a significant contribution to the teaching of science and biology in basic education.

REFERENCES

AMORIM, D. S. **Fundamentals of phylogenetic systematics**. Ribeirão Preto: Holos editora, 2002.

ARAÚJO-DE-ALMEIDA (Org). **Teaching zoology**: metadisciplinary essays, João Pessoa: Editora Universitária - UEPB, 2011.

BASTOS JÚNIOR, P. S. **Methodologies and strategies used to teach zoology.** 2013. 24 f. Monograph (Degree in Natural Sciences), University of Brasília, Planaltina - DF, 2013.

EMERICH, C. M. **Science teaching**: a proposal to adapt knowledge to everyday life - a focus on water. 2010. 156 f. Dissertation (Master's in Education). Federal University of Rio Grande do Sul, 2010.

GEHLEN, S. T. **The role of problems in the science teaching-learning process**: contributions from Freire and Vygotsky. 2009. 253 f. Thesis (Doctorate in Scientific and Technological Education) - Federal University of Santa Catarina, Florianópolis - RS, 2009.

HICKMAN, C. P.; ROBERTS, L. S.; LARSON, A. **Integrated Principies of Zoology**. McGraw-Hill Science, 2011.

LIKERT, R. A technique for the measurement of attitudes. **Archives of Psychology**. v. 22, n. 140, p. 44-53, 1932.

PEREIRA, N. B. **Perspectives for teaching zoology and possible directions for a different practice from the traditional one**. 2012. 43 f. Monograph (Degree in Biological Sciences), Mackenzie Presbyterian University - SP, 2012.

SILVA, C. D. D et al. Zoological caravan: contributions to science and biology teaching. In: National Congress of Education, 3, 2016. **Proceedings of the III National Congress of Education**, Natal: Realize Eventos e Editora, 2016.

CHAPTER 4

PRODUCTION OF COMIC STRIPS FOR THE PREVENTION OF DENGUE, ZIKA AND CHIKUNGUNYA

Naama Pegado Ferreira[14]

Ivaneide Alves Soares da Costa[15]

INTRODUCTION

Nowadays, it is undoubtedly necessary for teachers to be increasingly aware of the situations that occur in students' daily lives, so that they can teach in such a way that students can learn meaningfully. Among the many topics currently being addressed in the classroom is the Aedes aegypti and A. albopictus mosquitoes, which cause the diseases dengue, zika and chicungunya, which are emerging in our country and need to be prevented by everyone, as the Oswaldo Cruz Foundation (FIOCRUZ) warns in its infographic. "Vector control must be active and permanent, because different populations of A. aegypti and A. albopictus, which are highly capable of transmitting the Chicungunya virus, are distributed in Brazil," it also explains about the differentiation between the diseases, their prevention and the possible correlations with microcephaly in newborns and Guillain Barré Syndrome.

The school has become the best place to disseminate knowledge, especially in the prevention of diseases, as Donato (2005, p. 1) mentions:

> [...] in the case of these epidemics, outbreaks, pandemics or small concentrations of cases that are already a reason for the community and public health to be concerned, the school cannot be left out. But what is the school's role, then? It's to guide! That's what we do, we're counsellors. The school is responsible for adopting collective prevention measures. [...] On the basis that we can prevent infection. The school must inform itself about how to avoid diseases and pass on this information to its community of students, parents and educators. It is up to the school, whenever possible, to liaise with the public health sectors in their territories on preventative actions (DONATO, 2005, p.01).

With this in mind, there was a need to address this issue in classrooms in a different and attractive way. Among the textual genres that most attract the attention of schoolchildren and

14 Specialist in Science Teaching (IFRN) and Master's student in Natural Science Teaching (UFRN).
15 Master in Aquatic Bioecology (UFRN) and PhD in Ecology and Natural Resources (UFSCAR).

adolescents, Comic Books (Comics) stand out, as they co-operate both with the information needed for student learning and stimulate the imagination, as Calazans (2005, p.18) describes: "The different textual genres are capable of contributing in various ways to the construction of knowledge [...] from the development of analytical, interpretative and reflective capacity [...] to the stimulation of imagination and creativity".

The author goes on to say that: "the criteria for utilisation [...] must be defined and planned by the teacher, so that he or she can assess which type of comic will be most useful for teaching, based on the teaching and learning content and objectives [...] in his or her context of activity" (2005, p.21), and it is up to the teacher to adapt to his or her reality and define how this tool can be used to improve the teaching-learning process.

With this in mind, the aim of this project was to work with primary school students to produce comic strips in order to address the contents of prevention, symptoms, diseases caused by the A. aegypti and A. albopictus mosquitoes, and possible treatments, sensitising them to the importance of preventing these diseases.

METHODOLOGY

A total of 37 students took part in the activity, aged between 11 and 15, from the 6th year of primary school at the Castro Alves State School, Natal/RN. The activities lasted a total of 4 hours. Firstly, a short video was shown, produced by a group of students from USP (University of São Paulo), which deals with this subject in a dynamic way, covering everything from the emergence of the mosquito to the major concerns about the diseases caused by it. Later, through discussions, the importance of prevention and the difference between the symptoms of each disease were debated, as well as the microcephaly caused in newborns and Guillain Barré syndrome.

In the following lesson, a comic book produced and distributed by the Rio Grande do Norte State Health Department (SESAP/RN) was read, giving the students contact with the textual genre and information on the subject in question. The class was divided into groups (with up to 5 students) and they were asked to creatively produce their own comic book.

RESULTS AND DISCUSSIONS

After the video was shown, they discussed the different diseases and the damage they could cause, as well as ways of transmission and prevention. The discussion made it possible to check the students' prior knowledge. It was clear that the majority of the students had some knowledge of disease prevention, and they almost unanimously mentioned "eradicating the mosquitoes that cause them".

According to Oliveira, Martin and Bracht (2015), the school has been a place for health promotion actions, not least because of the recent Health at School Programme, with heavy investment in research and the development of government policies around health promotion due to the wide attention this issue has received, given its relevance. Schools are the right place for preventive actions such as these.

Despite this, many did not know how to differentiate between the signs and symptoms of the different diseases (Zika and Chicungunya), except for a few students who had already been affected by dengue, who were able to clearly describe the various symptoms they experienced during the period when they were affected by the disease. They were also asked about the reasons for the recurrence of dengue fever and whether the use of antibiotics would solve the problem. Many answered correctly and were able to correlate the use of antibiotics only with bacterial diseases, stating that their use would not be appropriate for treating dengue, as seen in previous lessons.

Brazil is a country with a high rate of television consumption by children, which is above the world average. They watch TV for more than 4 hours a day, with a great deal of participation in their daily lives, which is detrimental to their physical and mental development (ARAÚJO, 2016). Even with the media focus (newspapers, TV, internet, etc.) on this problem at an all-time high, none of the students knew what Guillain Barré Syndrome was before watching the video, but they did know about the possible correlation between the Zika virus and cases of microcephaly in newborns.

After the collective reading of the comic related to the theme, some questions arose about the difference between the viruses that cause illness and the mosquitoes that transmit the disease, and these were explained. Lino and Fusinato (2011) point out that this differentiation of concepts is a gradual process and that as meaningful learning takes shape, new concepts are assimilated, elaborated, developed and differentiated in the learner's cognitive structure.

The students were then divided into groups and instructed to create a story that was part of their daily lives and involved the prevention of one or more of the diseases they had studied. They were assessed on: formal writing in Portuguese, creativity in drawings, characters, organisation of the story and correlation with the correct fight against the mosquito. The students had a week to plan, design and present their work.

Group 01's production features a cover with two children holding a poster, which reads "Stop dengue" in English. The comic tells the story of Maria, a girl who was ill and had gone to hospital, discovering that she had dengue fever and returning home to clean the backyard. From this material, it is clear that the students correlate domestic cleaning as one of the activities to combat mosquitoes (Figure 01). Vaz et al, 2016, emphasise that health education is fundamental to the formation of individuals, their well-being and the improvement of the quality of life of the community, and since the school forms human habits, it is a vast field for the development of health promotion.

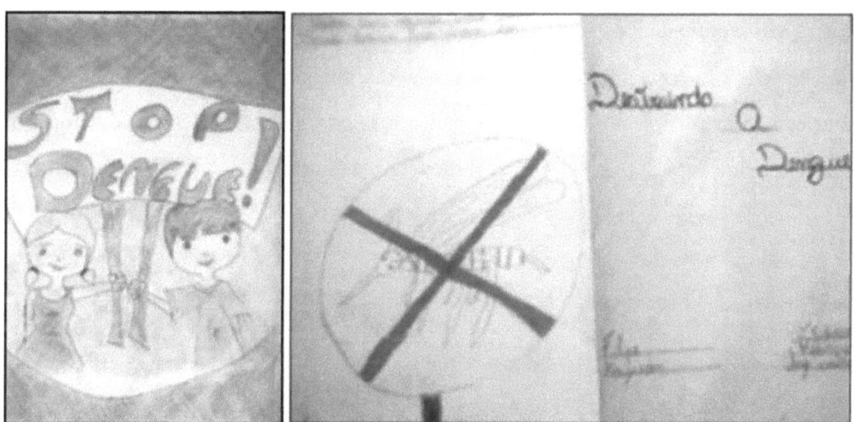

Figure 1: HQ cover and comic strips created by the students. **Source:** the authors (2017).

Group 02 drew up a story about children taking part in a community task force, teaching people how to fight dengue fever. The students specified some forms of prevention in the comic, such as: "don't leave standing water in tyres, bottles" etc. Group 03's theme was "destroying dengue" with a sign to combat the mosquito (Figure 02). In the story, two people were discussing what they could do to minimise mosquitoes, and during the conversation their idea was to "make posters and spread them around the city". The students didn't go back to their routine activities until after the work was done, and they also drew a computer with a mosquito on the screen being eliminated, which would also be a very effective form of prevention.

The theme of group 04 was "We're going to beat dengue" and the cover shows two boys playing in a backyard. There were bottles facing downwards and a sign showing a mosquito being eliminated on the wall of the house (Figure 04). The story had two characters, Rodrigo and Caio. One day, Caio fell ill and missed school. When his friend visited him, he explained how he avoided the mosquito: "don't accumulate water, don't leave dirty flower pots, cover water tanks, clean vacant lots, look after animal bowls". The comic ends with the friend saying "that doesn't sound difficult". This material shows that the students believe in the potential of preventive activities against these diseases, clearly demonstrating the real need for them.

According to Nercessian (1995), the thinking practices of scientists involve the construction of representations as well as other problem-solving activities, as was done in this practice, since the students proposed different ways of eliminating the mosquito. For her, the educator can be in a better position to establish strategies that facilitate the involvement of students in the practices of these scientists if they have a thorough knowledge of these tools.

Figure 2: HQ cover and comic strips created by the students. **Source:** the authors (2017).

Group 05 drew a mosquito on the cover and the title of the story was "The Zika virus". The material explained in detail how Zika is transmitted and how we can fight it. This was the group that cited the most ways to combat the mosquito. Group 06 told the story of three children who "caught dengue" from the mosquito's "bite". During the story, the children went to collect rubbish so that it wouldn't accumulate water, minimising/avoiding the development of mosquitoes. On the way, they found a lot of rubbish and didn't know where to put it.

In general, the material portrays a very common and worrying situation, raising a great deal of awareness about environmental awareness through the correct disposal of rubbish, and the development of preventive actions against the development of mosquitoes (Figure 05). Other studies, such as that by Barreto and Cunha, 2016, also indicate that the majority of students have a good understanding of recycling, which is considered relevant given that the improper disposal of rubbish has a major impact and negatively transforms the environment.

Group 07 also told a story about dengue fever. The names of the characters were the same as those of the group members. This shows that the students were included in the activity, as they saw themselves in a similar situation, and it was also justified by the possible occurrence of these diseases in the group members.

It turned out that although most of the students knew or referred mainly to the most common disease, dengue, this work contributed to significant learning on the part of the students, since they were evaluated not only for their creativity and participation in creating the story, but also for the possibility of learning and disseminating the knowledge needed to prevent these diseases, which have afflicted so much of the Brazilian population in the past year.

FINAL CONSIDERATIONS

We inferred that the students were able to learn about the topic and also put into practice in their daily lives ways of preventing related diseases by fighting the mosquitoes that cause them. It was also possible to verify their learning by assessing their reading and writing, their co-operation and their participation in the proposed activities. It is therefore suggested that interdisciplinary activities be proposed to boost the acquisition of skills and competences to improve reading, interpretation and writing and the commitment to exercising citizenship.

REFERENCES

ARAÚJO, V.E.S. **The influence of advertising on children's consumption: a motivating action in family consumption**. 2016. 60 f. (Course completion work for a bachelor's degree in Social Communication), University Centre of Brasília, 2016.

BARRETO, L.M.; CUNHA, J.S. Conceptions of the environment and environmental education by primary schools students in Cruz das Almas (BA): a case study. **Brazilian Journal of Environmental Education**, São Paulo, v. 11, n. 1, 2016. p. 315-326.

CALAZANS, F. **História em quadrinhos na escola**. São Paulo: Paulus, 2005.

DONATO, A. F. **The role of the school in disease prevention**. Available at: <http://educacao.estadao.com.br/blogs/colegio-equipe/o-papel-da-escola-na- prevencao-de-doencas/>. Accessed on: 25 April 2016.

LINO, A; FUSINATO, P.A. The influence of prior knowledge in the teaching of Modern and Contemporary Physics: an account of conceptual change as a process of meaningful learning. **Revista Brasileira de Ensino de Ciência e Tecnologia**, Ponta Grossa, v.4, n.3, Sept - Dec. 2011.

NERDOLOGY. **Zika virus.** Available at: <https://www.youtube.com/watch?v=pm3do0nEuuM>. Accessed on: 23 March 2016.

NERSESSIAN, N. Should physicists preach what they practice? Constructive modelling in doing and learning physics. **Science & Education**, v.4, n.1, 1995. p. 203-226.

OLIVEIRA, V.J.M; MARTINS, I.R; BRACHT, V. Relations between physical education and the school health programme: views of teachers from schools in Vitória/ES. **Thinking about Practice**. Goiânia, v. 18, n.3, jul. - sep. 2015. p. 544 - 556.

VAZ, A. V.M.; FERRAZ, G.T.; MASSARDI, F.R.; CLÍMACO, S.S.; MARTINS, Y.P.; RODRIGUES, L.P. How can we teach body hygiene beyond the classroom? Experience Report. **Journal of Management and Primary Health Care**. Recife, v.7, n. 1, 2016. p.83-83

CHAPTER 5

FUN ACTIVITY FOR TEACHING PHOTOSYNTHESIS IN SECONDARY SCHOOL

Naama Pegado Ferreira[16]
Ivaneide Alves Soares da Costa[17]

INTRODUCTION

Studies show that the teaching and learning of biological processes is one of the most challenging because the content is abstract and complex (HAAMBOKOMA, 2007, BOUJEMAA et al., 2010). One of these is the photosynthetic process, which is considered very important because it is responsible for the input of energy and the maintenance of life on planet Earth (ROSS et al., 2006). According to Kawasaki and Bizzo (1999, p. 28): "in science teaching, photosynthesis should not be approached as an isolated topic, but in the context of the processes that realise autotrophic nutrition". For this reason, the teacher should look for didactic alternatives to contextualise and make understanding the concepts easier and more attractive.

In this sense, playful activities offer the advantage of making lessons less tiring and stimulating, making teaching more enjoyable and attractive (KETCHEL; BRANCALHÃO, 2008), as well as enhancing learning in a more meaningful way. According to Campos, Bortollo and Felício (2008), the game, as a playful activity, is an ideal learning tool, in that it stimulates the student's interest; develops different levels of personal and social experience; helps them build their new discoveries; develops and enriches their personality and symbolises a pedagogical tool that takes the teacher to the position of conductor, stimulator and evaluator of learning.

The aim of this study was to help first-year secondary school students learn about the photosynthetic process.

16 Specialist in Science Teaching (IFRN) and Master's student in Natural Science Teaching (UFRN).
17 Master in Aquatic Bioecology (UFRN) and PhD in Ecology and Natural Resources (UFSCAR).

METHODOLOGY

As a way of consolidating the content on photosynthesis, a didactic game (Figure 1) was developed and applied, lasting two 50-minute lessons, to 72 students from two 1st year high school classes at a public school in the municipality of Natal/RN.

The game begins with the participants in each group (with up to 5 components) dividing into opposing subgroups (A and B). The game has a "judge" who keeps the questions in hand and organises the answers for each subgroup (Figure 2).

Out of the 05 components, 01 is selected to be the "judge", leaving 02 students from sub-group "A" and 02 from sub-group "B" to compete in the game. One of the members of each subgroup draws an odd or even number to start the game. The winning student has the right to throw the dice and start the game.

During the game, each student has up to one minute to answer the questions. For every wrong answer, they have to go back to the beginning. The winning team is the sub-group that enters the finishing square (the sheet).

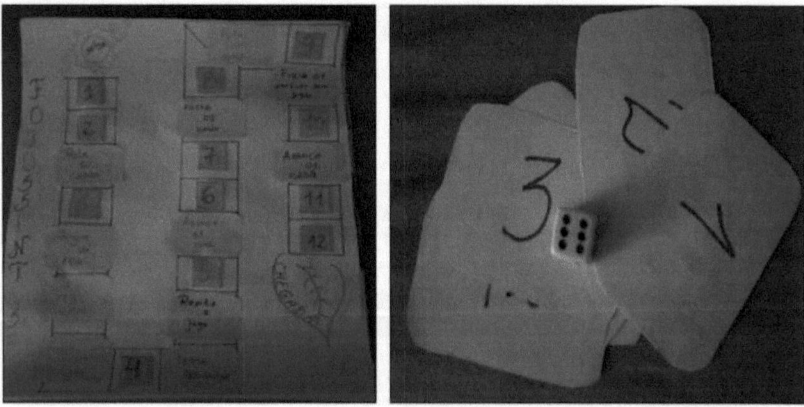

Figure 1: Educational game designed to address photosynthetic processes. **Source:** the authors (2017).

Figure 2: Judge. Subgroups "A" and "B" during the application of the didactic game. Source: the authors (2017).

RESULTS AND DISCUSSIONS

In the first class, some of the groups who had mastered the content had no difficulty in playing the game, participating in more than one game during the class, which favoured the score of each subgroup. As the questions were repeated throughout the games, the other components had the opportunity to learn and answer, moving on to the next questions.

According to Frantz (2001, p.243): "[...] education and co-operation are two social practices which [...] in certain respects, one contains the other. Education is a fundamental social process in human life. In cooperation as a social process, education is produced," in other words, they are intertwined and potentialised. It is from this perspective that this game, despite instigating competition, has traces of co-operation between the non-competing groups.

The photosynthesis content covered in the game was: light and dark phases of photosynthesis, water photolysis, the organelle responsible for the process in plants, the functions of NADP and ATP molecules, photophosphorylation and the Calvin cycle.

With regard to the rules of the game, it's important that the teacher, as a helper in this process, changes the rules only if necessary. If one group finishes first, for example, the teacher can encourage them to play again by competing to see how many times they have reached the end. This situation occurred in

groups in which they had mastery and knowledge of the subject. Anastasiou, 1998 states that learning can take place in a variety of ways, depending on the subject and the object, which respectively apprehends and is apprehended. One of the ways of learning cited is by imitating a model, by repetition, by trial and error, by discovery, among many others. As such, the repetition of the game as a form of learning is considered positive.

Among the students, it was possible to see that approximately 20 per cent of the total number of participating students did not have a grasp of the aforementioned content, but were able to learn it during the development of the game. They played more slowly and, whenever necessary, asked for the teacher's help.

The majority of students (80%) demonstrated mastery of the questions presented on the photosynthetic phases, organelles and photolysis of water. Despite this, the students had difficulties with the Kelvin Cycle and the molecules involved in the photosynthesis process. Respect for each student's learning time is important: "[...] humanistic theory is characterised by respect, trust, acceptance and tolerance of each student's pace of learning. Thus the role of the teacher goes beyond teaching to facilitating student learning." (JUNQUEIRA; SILVA, 2016, p.94).

In all the classes, the students were enthusiastic about the game and expressed their positive opinions and appreciation for the dynamic. We believe that this situation is possibly related to the ease of learning through play. Souza (1996, p.5) emphasises that:

> Play is a valued dimension in the development of learning, encompassing children and adults, taking them to even greater heights through play and representing the need to know, build and relax in a real or symbolic world full of marvellous moments that only happen through play. (SOUZA, 1996, p.5)

At the end of the application, comments such as: "I thought today's lesson was great", "I wish there were more of these" were common, highlighting the importance of playful activities compared to the practices commonly seen in the classroom, as these favour and stimulate student learning.

FINAL CONSIDERATIONS

In conclusion, this is a dynamic way of encouraging student participation in the teaching and learning process. Teachers can use it to continually assess students, as it is recommended that teachers are continually checking the progress of student learning.

It was possible to ascertain the students' learning over the course of the game through the correct answers to the questions they asked in order to progress in the game. This provided evidence of the game's contribution to learning about the subject.

In view of this, it is suggested that other educational products be created, preferably with rules created and/or adjusted by the students themselves, which are low-cost and provide more dynamism during science and/or biology lessons.

REFERENCES

ALMEIDA, P. N. **Educação Lúdica**. 9 ed. Rio de Janeiro, Loyola, 1992.

ANASTASIOU, L. G. C. **Metodologia do Ensino Superior**: da prática docente a uma possível teoria pedagógica. Editora IBPEX, Curitiba, 1998.

BOUJEMAA, A. et al. University students' conceptions about the concept of gene: Interest of historical approach. **US-China Education Review**, v.7, n.2, p.9-15, 2010.

CAMPOS, L. M. L; BORTOLOTO, T. M.; FELICIO, A. K. C. **A produção de jogos didáticos para o ensino de ciências e biologia**, 2008. Available at: <www.une sp. br/prograd/PDFNE2002/aproducaodej ogos. pd>. Accessed on: 15/03/2017.

FRANT, W. Education and co-operation: practices that relate. **Sociologias**, Porto Alegre, v. 3, n. 6, july - dec. 2001. p.242 -264.

HAAMBOKOMA, C. Nature and causes of learning difficulties in genetics at high school level in Zambia. **Journal of International Development and Cooperation**, v. 13, n. 1, p.1-9, 2007.

KAWASAKI, C. S. **Plant nutrition: a field of study for science teaching.** 1998. 134 f. Thesis (Doctorate in plant production), University of São Paulo, SP, 1998.

KNECHTEL, C.; BRANCALHÃO, R. **Playful Strategies in Science Teaching.** 2008. Available: <http://www.diaadiaeducacao.pr.gov.br/portals/pde/arquivos/2354- 8.pdf>. Accessed on: 15/04/2017.

SANTOS, A. O.; JUNQUEIRA, A. M .R.; SILVA, G. N. Affectivity in the teaching and learning process: dialogues in Wallon and Vygotyski. **Perspectives in Psychology**. v. 20, n.1, p. 86-101, 2016.

SOUZA,C. F. **The importance of children's play and learning in early childhood education**. Available at: <http://facsaopaulo.edu.br/media/files/58/58_161.pdf>. Accessed at: 25/05/2017.

TOPÇU, M. S.; &AHYN-PEKMEZ; E. Turkish Middle School Students' Difficulties in learning Genetics Concepts. **Journal of Turkish Science Education**, v. 6, n. 2, p. 55-62, 2009

TOPÇU, M. S.; &AHYN-PEKMEZ; E. Turkish Middle School Students' Difficulties in learning Genetics Concepts. **Journal of Turkish Science Education**, v. 6, n. 2, p. 55-62, 2009

CHAPTER 6

DIDACTIC MODELS IN THE TEACHING OF CELL BIOLOGY: A PROPOSAL TO HIGHLIGHT AND OVERCOME ALTERNATIVE CONCEPTIONS

Naama Pegado Ferreira[18]

Clécio Danilo Dias da Silva[19]

Ivaneide Alves Soares da Costa[20]

INTRODUCTION

It is currently essential in science teaching to study cells (cytology), working on their concepts, morphology, structures and functions so that students can have quality scientific literacy while still in basic education.

It is well known that a large number of state schools lack equipment, for example specialised laboratories containing glassware, magnifying glasses, microscopes and other essential equipment for better visualisation and understanding of the study of cell biology. Thus, didactic models stand out as one of the most viable alternatives for teaching and illustrating science lessons, particularly the study of the cell.

In this way, the use of didactic models is relevant when teaching this subject, as they are an alternative tool that helps to minimise possible learning difficulties on the part of students. Amaral (2010) mentions in his work that the handling of a three-dimensional cell model by students provides great interest and curiosity. This playful way of learning brings students closer to scientific concepts in a pleasurable and meaningful way.

In this sense, considering that the content of cytology is difficult for students to learn, we realised the importance of encouraging studies in this area in secondary schools, proposing research and developing teaching models.

18 Specialist in Science Teaching (IFRN) and Master's student in Natural Science Teaching (UFRN).
19 Specialist in Science Teaching (IFRN) and Master's student in Natural Science Teaching (UFRN).
20 Master in Aquatic Bioecology (UFRN) and PhD in Ecology and Natural Resources (UFSCAR).

METHODOLOGY

The activities were carried out with two 1st year high school classes at the Castro Alves State School, Nova Descoberta, Natal - RN. Initially, the classes were divided into groups (with up to 6 members) and asked to make 3 prototype cells (prokaryotes, animal eukaryotes and plant eukaryotes), showing their organelles.

They were also asked to use low-cost materials to make the models, preferably recyclable and/or reusable materials (cardboard, cans, pet bottles, etc.). The students were informed that they would be assessed on their creativity and mastery of the content during the presentation.

The students were free to do their research in various sources of information (textbooks, magazines, the internet), looking for information about cells to make their models. A period of 15 days was set aside for the students to make and socialise their prototypes with the class. On the day of the socialisation, the teacher approached and questioned the groups whenever necessary. After the presentations, the models were displayed in the school's science laboratory.

RESULTS AND DISCUSSIONS

Through the models produced and socialised, some conceptual errors and alternative conceptions on the part of the students became apparent, especially with regard to the dimensions of the organelles.

Similar work also demonstrates the relevance of modelling and the difficulty in grasping certain concepts and cell dimensions, as Elias, Siqueira and Santos (2016, p. 2) cite: "[...] students find it easier to learn by handling these models, as there is great difficulty in abstraction on the part of students when the content taught is not something concrete, tangible for them."

As for the differentiation of cells, the groups were able to identify them more easily, including the organelles that each one has, with their respective functions, since they had made the models themselves.

As for prokaryote cells, because they are simpler and have fewer organelles, several groups used recycled materials to represent them. The prototypes were mostly made from recyclable materials (PET bottles, disposable straws, paper) for the outside of the cells, plus low-cost materials (modelling clay, paint, toothpicks) for the inside. The students placed cilia and flagella (structures of some prokaryotic cells), contributing to learning about cell locomotion. These structures were also used as questions in the didactic game, Cell Bingo, in the work by Gonçalves et al. (2014), showing the relevance of learning about them.

Some groups did not show the locomotion structures, but this is not a conceptual error, since not all prokaryote cells have these structures (Figure 01A). In the models in the figure below, the ribosomes, genetic material, absence of an organised nucleus, cell wall and plasma membrane were evident. One of the groups made their model out of clay, modelling clay and string (flagella), which is more expensive but more durable (Figure 01B).

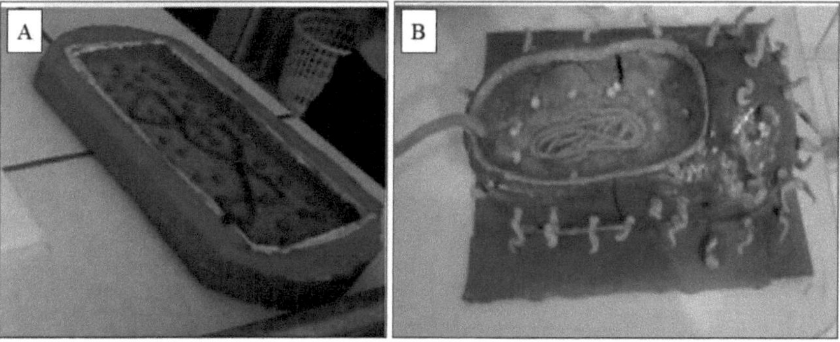

Figure 1: Model of a prokaryote cell without a flagellum (A), model made with clay and modelling clay.
Source: the authors (2017).

In the models representing the eukaryotic plant cell, some groups used materials such as styrofoam, cardboard, modelling clay, plastic lids and tablets (Figure 02A and 02B). There was a marked presence of the plant cell wall, vacuole, organised nucleus, Golgi complex, granular and non-granular endoplasmic reticulum.

The students also clearly represented the presence of microfilaments, which are part of the cell structure. Although the students' concern to illustrate them as a cellular component is relevant, care must be taken not to induce the construction of alternative conceptions, since these structures are much smaller than organelles. We would also highlight the inadequate position of the endoplasmic reticulum, which does not visually portray its transport function. These aspects were discussed in class with the students. In order to avoid possible conceptual errors, the concept of the cell needs to be clear to the student, which is not always possible using the textbook alone, according to the research by Silva, Nocelli and Bozzini (2015, p.33):

> The concept of the cell, although it is not dealt with in a direct way, with descriptions, not presenting a definition, provides support so that the student can understand the importance of cells for life and how it relates to the functioning of the body as a whole [...] To make it clearer what a cell is, it needs to be approached in a different way, it must be contextualised through its history, relevance and demonstrated its importance for the origin of life, so that the student understands its role in a more concrete way, so that he can form his conceptions about it.

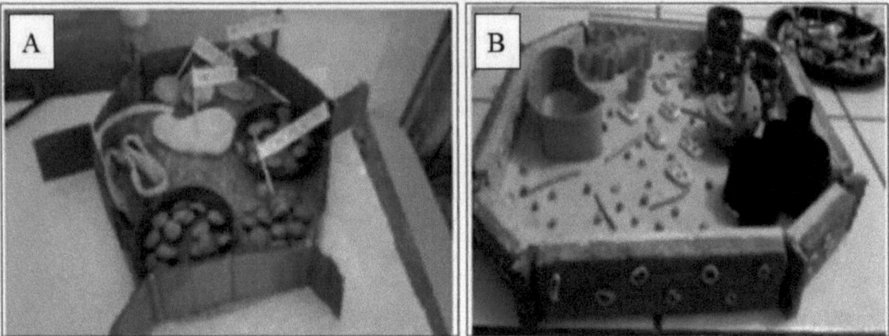

Figure 2: Model of a eukaryotic plant cell developed with (A) cardboard and (B) styrofoam. **Source:** the authors (2017).

Another group chose to make their model of a plant cell out of a cake. The students used American paste in different colours to represent the organelles. Since it is an edible material, this was not a durable model, but it was creative. Morais, Azevedo and Jesus (2014) in their studies emphasise that both creativity and motivation are intrinsically related, especially in the context of school learning, which makes it clear that students need to use creativity as a teaching strategy.

To develop the models representing the animal eukaryote cell, the students used various materials, such as cake, artificial sponge, modelling clay, styrofoam balls, disposable plates and cardboard (Figure 03A, B and C). In the model shown in Figure 03A, the organelles were visible "immersed" in the cytosol, just as they are in the cell. This group represented the organelles with modelling clay and coloured E.V.A., respecting the different shapes of each one. They used transparent gel for the cytosol.

The group that chose to represent the cells using edible material (Figure 03B) did not show the differentiation of the organelles, only the nuclear organisation, which is distinct from the prokaryote cell. Another group of students developed their model with low-cost resources, using only coloured modelling clay and a disposable plate (Figure 03C). In this model, despite its simplicity, the different organelles and the presence of microfilaments were noticeable (although the dimensions did not conform to a standard).

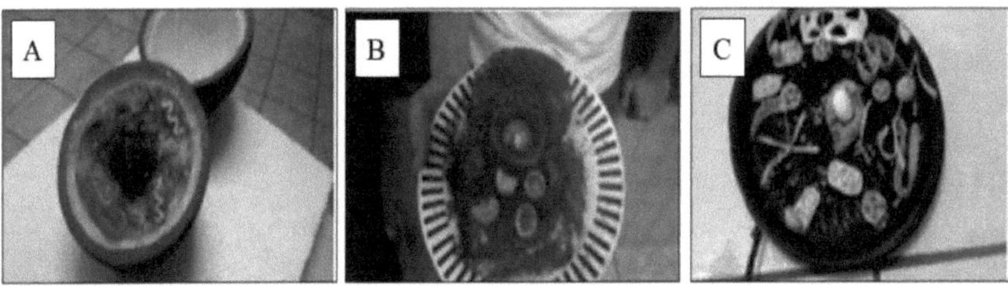

Figure 3: Animal eukaryote cell model developed with (A) styrofoam, modelling clay and EVA, (B) edible materials, (C) modelling clay and disposable plate.
Source: the authors (2017).

There were groups of students who, because they considered the animal eukaryote cell to be more important than the others (possibly because it has more structures and organelles), only developed a representation of an animal cell. In view of this, we observed that these students had difficulties with their knowledge of the differentiation and dimensions of the organelles, as well as the distinction between the other cells. Work like this corroborates the evidence found in previous studies, such as that by Vinholi Junior and Gobara (2016), in which these modelling tools and/or concept maps are potentially significant tools for teaching cytology.

One of the groups presented their cell prototype in a sponge (Figure 4A). In it, the organelles were represented in modelling clay, showing the plasma membrane, cytosol, mitochondria, ribosomes, endoplasmic reticulum and nucleus. Although the material was well prepared, most of the students did not know how to differentiate between the organelles or their functions. Another group of students, using the prepared material (Figure 4B), made some conceptual errors, as they believed that ribosomes were stuck/immobile in the cytosol. The model also showed the absence of some important organelles. The students were unable to differentiate between the types of cells, but they did have knowledge of the names and respective functions of the organelles.There were groups that had more difficulty assembling the prototype than learning the content (Figure 4C), since they wanted to develop a nucleus that could move around inside the cell using a barbecue stick. The objective was achieved, but it disrupted the organisation and dimensions of the cell. The three groups had in common the absence of important organelles for the cell, as well as a legend on their prototypes, making it easier to identify and associate the different organelles with their functions.

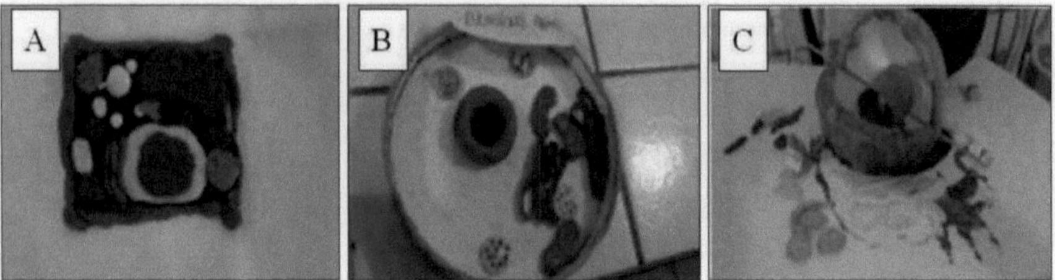

Figure 4: Model of an animal eukaryote cell (A) made with a sponge (B) lacking several organelles (C) with a mobile nucleus.
Source: the authors (2017).

As the groups socialised their models, the teacher questioned them about the materials they had made, the cost of making them, and the difficulties and facilities they encountered. The students were also asked about the differences between the cells. During the presentation, some of the students were shy, while others were more adept at oral presentation. There was a consensus in most of the groups that all the participants who were present had done the work.

For Barbosa and Wiezzel (2015), shyness can also affect creativity, which is not interesting for students, especially with the type of activity proposed, which required creativity and resourcefulness in the presentation, in order to better highlight their previous knowledge.

Through the presentations and answers to the questions, it was possible to gradually realise the students' alternative conceptions and conceptual errors. However, after socialising the materials, the students were able to compare the structures of their models with those of the other groups, identifying and overcoming their conceptions about the structures of organelles and the distinction between cells. We emphasise that this event was important for checking the degree of interest and progression of the students in learning the content covered, as well as verifying the difficulties each had, and the presence and/or absence of possible alternative conceptions.

FINAL CONSIDERATIONS

In view of the findings, this work was a valid way of assessing students' alternative conceptions of cytology, helping them to identify and overcome them. In addition, the construction of the models contributed to learning about animal and plant cell structure. The students were able to demonstrate the knowledge they had acquired in a different way, building their own knowledge. This work can also be used to highlight the importance of reusing and recycling rubbish, as some of the models were produced using disposable, low-cost materials.

REFERENCES

AMARAL, S. R.; COSTA, F. G. **Strategies for teaching science:** Three-dimensional models - a new approach to teaching the concept of the cell. State University of Maringá. Available at: <www.diaadiaeducacao.pr.gov.br/portals/pde/arquivos/1864-8.pdf >. Accessed on: 30/04/2017.

BARBOSA, A.C.V.;WIEZZEL, A.C.S. Shyness in early childhood education: contribution and intervention through play. In: UNESP University Extension Congress,8, 2015. **Proceedings**... São Paulo: UNESP, 2015.

ELIAS, F.G.E.; SIQUEIRA, P.A.; SANTOS, M.L. APPLICATION AND EVALUATION OF DIDACTIC MODELS: A PROPOSAL FOR TEACHING CELLS IN PRIMARY SCHOOL. In: UEG Teaching, Research and Extension Congress, 3, 2016. **Proceedings**... Pirenópolis: State University of Goiás, 2016.

GONÇALVES, R.R. et al. Cell bingo: a methodological tool for teaching cell biology. **Revista Ensino & Pesquisa**, Campo Mourão, v.12, n.01, 2014. p. 28 - 47.

HAZEN, R.; TREFIL, J. **Knowing Science from the Big Bang to Genetic Engineering the Basics for Understanding the World Today and What's Next.** Translation by Cecília Prada. São Paulo. Editora de Cultura, 2005.

LOPES, S. **Bio 1**. São Paulo: Ed. Saraiva, 2012.

MORAIS, M. F.M.; AZEVEDO, I.; JESUS, S. Creative competences and motivation for learning: Different realities in adolescents? **Revista de Psicologia, Educação e Cultura**, Porto, v.18, n.1, 2014. p.87-99.

ORLANDO, T. C. et al. Planning, assembly and application of didactic models for approaching cellular and molecular biology in high school by undergraduate students of biological sciences. **Revista brasileira de ensino de bioquímica e biologia molecular**, v. 01, n.1, p. 1-15, 2009.

SILVA, L.C.D.; NOCELLI, R.C.F.; BOZZINI, I. C.T. Approach to cells in teaching materials. **Science, Technology & Environment Journal.** Araras, v. 1, n. 1, 2015. p.28-34.

SILVA, M. G. L.; SILVA, A. F.; NÚNEZ, I. B. From conceptual change models to learning as guided research. In: NUNEZ; I. B.; RAMALHO, B. L. (Orgs.). **Fundamentals of teaching and learning the natural sciences and maths: the new secondary education.** Porto Alegre: Sulina, 2004.

VINHOLI JÚNIOR, A.J.; GOBARA, S.T. Teaching with models as a tool to facilitate learning in Cell Biology. **Revista Electrónica de Ensenanza de las Ciencias**, Vigo , v. 15, n. 3, 2016. p. 450-4

CHAPTER 7

PLAY TO ENCOURAGE READING AND ASSESS LEARNING IN SCIENCE TEACHING

Naama Pegado Ferreira[21]

Ivaneide Alves Soares da Costa[22]

INTRODUCTION

Science teaching is undoubtedly very relevant and thought-provoking, as it leads students to reason, create their own thoughts and decisions, providing them with intellectual autonomy, as Paulo Freire proposes in his vast work (19211997). However, it is often up to the teacher to create different ways of achieving these goals, by being a moderator of student learning, motivating them to always be co-operating with other students and encouraging them throughout the teaching and learning process. According to Jean Piaget *apud* Pereira (2008):

> It's up to the teacher to propose problems without teaching solutions; to propose challenges, to provoke imbalances. The teacher's role is that of investigator, advisor, researcher and coordinator, and the student's role is that of independent subject, working by observing, experimenting, relating, comparing, composing, juxtaposing, hypothesising, arguing, etc. (Piaget *apud* Pereira, p. 15).

One of the ways teachers can fulfil their arduous role, often with few resources available and in a way that appeals to students, is through playful teaching. According to Knechtel and Brancalhão (2008, p.02): "we can say that in playful activities we go beyond reality, transforming it through imagination [...] it can be used as a promoter of learning in school practices, making it possible to bring students closer to scientific knowledge

Play also sharpens curiosity and imagination, which is very important for students'

21 Specialist in Science Teaching (IFRN) and Master's student in Natural Science Teaching (UFRN).
22 Master in Aquatic Bioecology (UFRN) and PhD in Ecology and Natural Resources (UFSCAR).

education, especially when it comes to learning the natural sciences, because: "The exercise of curiosity summons the imagination, intuition, emotions, the ability to conjecture, to compare, in the search to profile the object or to find its reason for being". (FREIRE, 1996, p. 98)

With this in mind, this study was carried out during science lessons with students in the 6th year of primary school at a public school in the municipality of Natal, who have learning difficulties and/or low levels of academic achievement. These students often end up failing or dropping out of school. From this perspective, we realised the need to consolidate and recur previously worked content in a dynamic way, actively involving participation and cooperation among the students, encouraging reading and, consequently, writing.

METHODOLOGY

The dynamics lasted just one hour, and 30 questions were selected in advance, referring to the content being taught during the second two months of the school year in Science, such as: food chain and food web, autotrophs, heterotrophs, producers, consumers and decomposers. Among the questions there were 03 extra points that were randomly placed between them.

The students were divided into two groups (boys and girls). The challenge proposed to everyone was that the group that got the most questions right would receive gifts (books) (Figure 01). It's worth pointing out that all the children's books had been donated to the school by an institution that was shutting down its library. A volunteer from each team was asked to start the game by drawing odd or even with the competing group. Once this was done, they were given the opportunity to start the game. The rules previously presented were that the students in each group could consult their notes and textbooks to answer the questions, but would only have 40 seconds to answer correctly. After this period, the chance was passed on to the next group. When the allotted time was up, the teacher would answer the question if no group got it right, clearing up any doubts.

Figure 01 - Paradidactic books and magazines donated to the school.
Source: The authors (2017).

Each participant in the groups was responsible for choosing a number, and they could score an extra point (without the need for questions) if they removed the number corresponding to this item (Appendix 01) or removed a question, which had to be answered. The numbers that had already been chosen could not be repeated, as the questions had already been prepared in advance, as mentioned above. The score was gradually recorded on the board according to the errors and successes of the groups, which further encouraged the fulfilment of the stipulated time and the search for answers and learning memories. (Figure 02)

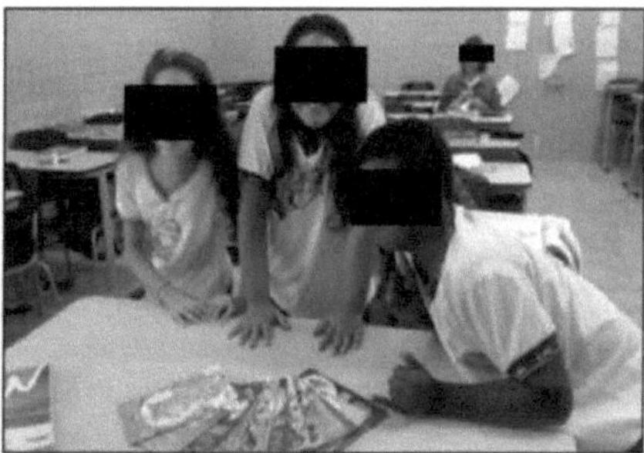

Figure 02 - Scoring and selection of booklets by the students. Source: The authors (2017).

RESULTS AND DISCUSSIONS

During the competition, the students were assessed on their mastery of the content, as well as their co-operation in answering questions, resolving doubts and learning difficulties. (learning competitions quote)

The activities initially generated a great deal of euphoria on the part of the students, but they gradually adapted, respecting the rules and trying to answer the questions that were asked throughout the game, being able to recall the content previously taught and discussed in class.

Rodrigues (2017, p.36) argues that: "What changes or does not change the quality of learning assessment is teaching. Thus, it is believed that we need to change both the ways of assessing and the ways of teaching". This dynamic is an educational way of doing both things, both teaching and assessing, since they are continuous processes.

After this activity, the co-operation between the participants in the same group was noticeable, despite it being a "competition", they were also able to be assessed in a natural way without the pressure of a written test, which also increased the performance of some students in the bimonthly test, with a percentage of 30 % more correct answers.

Nowadays, there are various ways of continuously assessing students without the "pressure" of a written test, as suggested by Vieira (2002), in which the use of portfolios by students cooperated in their assessment, and there was also a higher than expected academic performance, with no student below the required average. We can therefore consider that all these different forms of assessment from the traditional ones are, for the most part, also positive for student learning.

You can also see the students' excitement at their achievement, even when they consulted the textbook, which was one of the biggest incentives in their search for answers and also helped in their learning process.

FINAL CONSIDERATIONS

It is clear that with creativity and playfulness, teachers can work to improve education, even at no financial cost, by encouraging students to take part in lessons, promoting playful activities related to the teaching of science or another subject.

It is also concluded that the incentive given through the distribution of paradidactic books was of great value to the students in consolidating their learning of the science content studied.

It is understood that the groups could be divided randomly, not just by gender, forming larger and/or heterogeneous groups, depending on the number of students and each reality. It is also suggested that these playful activities be carried out in spaces other than the formal teaching spaces, "escaping" the daily routine of the school classroom, promoting incentives for learning and improving teaching in the public network.

REFERENCES

FREIRE, P. **Pedagogia da Autonomia**: Saberes necessários à prática educativa. São Paulo: Paz e Terra, 1996.

KNECHTEL, C. BRANCALHÃO, R. **Playful Strategies in Science Teaching.** 2008. Available at: <http://www.diaadiaeducacao.pr.gov.br/ /2354-8.pdf>. Accessed on: 15/04/2017.

PEREIRA, M. A. **A Importância do Ensino de Ciências: Significant Learning in Overcoming School Failure,** SEED, 2008. Available at:< http://www.diaadiaeducacao.pr.gov.br/portals/pde/arquivos/2233-8.pdf>. Access:01/042017.

RODRIGUES, F. C. A.. Learning assessment and assessment practices in youth and adult education (EJA): reviewing myths, rites and realities. **Debates in Education**, v. 9, n. 17, p. 35-49, 2017.

VIEIRA, V.M.O. Portfólio: uma proposta de avaliação como reconstrução do processo de aprendizagem. **Psicologia Escolar e Educacional**, Uberlândia, v.6, n. 2, 2002. p.149-153.

Appendix 01 - Sample game questions

01 - What is an ecosystem?
02 - Give an example of a food chain
03 - What is the difference between producer, consumer and decomposer?
04 - **Extra Point**
05 - In the given food chain, which is the tertiary consumer?
06 - Who are the decomposers?
07 - Differentiate between food webs and food chains.
08 - Give an example of heterotrophic beings;
09 - What are autotrophic beings?
10 - Plants produce their own food. What are they called because of this characteristic?

CHAPTER 8

THE USE OF PEDAGOGICAL WORKSHOPS AS A DIDACTIC RESOURCE IN SEXUALITY EDUCATION

Clécio Danilo Dias da Silva[23]

Carmem Maria da Rocha Fernandes[24]

Daniele Bezerra dos Santos[25]

Lúcia Maria de Almeida[26]

INTRODUCTION

Human sexuality is currently recognised as an important aspect of people's health and quality of life. However, despite the sexual revolution, globalisation and the media, this subject is still taboo. Often, for reasons of gender, religion, lack of information or even not knowing how to approach the subject, parents or carers don't feel comfortable talking about it. In this context, the school is often left with the responsibility of working on the issue, especially given the high rate of teenage pregnancy, which is one of the causes of high school dropout rates.

Schools often don't have a plan or planned actions in the school environment that can offer guidance to adolescents and parents in the area of sexuality. Most of the time, teachers and schools are not prepared to deal with issues relating to sexuality, limiting themselves to emphasising only the biological and physiological conception of the subject. According to the National Curriculum Parameters (BRASIL, 1998), the approach to sexual orientation at school should be carried out systematically, providing critical, reflective and educational action that promotes the health of children and adolescents.

In this context, the aim of this work was to stimulate reflection on the knowledge of sexually transmitted diseases that make it difficult to establish preventive attitudes and to increase

---

23 Specialist in Science Teaching (IFRN) and Master's student in Natural Science Teaching (UFRN).
24 Specialist in Teaching Natural Sciences and Mathematics (IFRN).
25 Master in Bioecology (UFRN) and PhD in Psychobiology (UFRN).
26 Master in Botany (UFRPE) and PhD in Psychobiology (UFRN).

participants' knowledge and skills in preventing unwanted/unplanned pregnancies by reducing their exposure to the risks of abortion.

METHODOLOGY

The activities were carried out with students in the 7th year of primary school II, aged between 14 and 17, at the José Fernandes Machado State School, in the municipality of Natal - RN. Initially, a meeting was held with the school management and pedagogical coordination, in which the choice of classes was discussed according to the need to address the issue of prevention of sexually transmitted diseases and unplanned pregnancy. The meetings with the students took place between August and November 2015, during science lessons, totalling five meetings per class (Figure 1).

Figure 01 - Awareness-raising, question box, display of the morphophysiology of the reproductive organs, survey of known CMs and simulation of the STD contamination network.

Source: The authors (2017).

In the first meeting, we carried out the sensitisation stage through a round table discussion. The second meeting dealt with the questions they asked via the question box, as well as debates on pertinent subjects, mainly on how to avoid unwanted pregnancies and the correct use of condoms. At the third meeting, we held an exhibition with anatomical models describing basic aspects of sexual and reproductive anatomy and physiology, differentiating between the structures and functions in girls and boys.

In the fourth meeting, a survey was carried out on the contraceptive methods known by the adolescents, emphasising the practice of safe sex as the most effective way of preventing STDs and unplanned pregnancies. In the fifth meeting we simulated a network of contamination using different coloured candies, where a particular colour could represent an STD, and systematised the stages of correct condom use using cards describing the stages without sequential order, which were given to them to place on a sequentially numbered ladder.

RESULTS AND DISCUSSIONS

At the pedagogical meeting with the school's management and coordinators, it was requested that the workshops be developed in the respective 7th grade classes B and C. They had as references many cases of unplanned pregnancy by teenagers and several reports of students with sexually transmitted diseases. According to the coordinator, the teachers didn't feel prepared to tackle the issue because they didn't have access to a theoretical framework or activities on the subject.

According to Moizés and Bueno (2010), the teacher does not need to be a specialist in sex education, but just a professional who is properly informed about human sexuality and who reflects on it, being able to create appropriate pedagogical contexts, as well as selecting methodological strategies that allow for reflection and debate of ideas, in addition to information, and for this it is necessary for them to take part in continuing training that allows them to update their knowledge.

In an effort to create an atmosphere in which the students felt comfortable talking about the subject, we began the first meeting with a round table discussion, with the aim of sensitising them to take part. It was possible to realise their existing knowledge of the subject, as well as to provoke adolescents to think about caring for their bodies. According to Lima and Pagan (2010), offering a space for dialogue, based on welcoming doubts and providing direct information that is appropriate to the adolescent's development and context, seems to be the best way to approach the subject of sexuality.

We realised that some of the students didn't feel comfortable talking more openly about the subject, so we used the strategy of placing questions in a box to be discussed at the next meeting. We realised from the questions asked anonymously in the box that there was a negative attitude towards sexuality, often related to something dirty, ugly or sinful. The majority of young people and teenagers don't know or know the names of the organs that make up the male and female reproductive system, nor do they relate the organ to its respective function; in addition, it is quite common for the organs to be used or known by nicknames.

The survey on contraceptive methods indicated that practically everyone knows about or uses condoms, but when we looked at how to use them correctly, the vast majority of young people (boys and girls) didn't know how to use them properly. Many didn't bother to check the packaging, whether it was damaged, the expiry date, or tear the packaging with their teeth, one of the most frequent mistakes they made. We also realised that many adolescents and young people believe that they are immune to sexually transmitted diseases, including AIDS. Many girls believe that when they have sex for the first time they won't get pregnant.

The teenagers were keen to take part in the discussions and the stages of the activities developed at each meeting. We realised that pedagogical workshops are a teaching strategy that allows a problem situation to be analysed, as well as reflection in the search for a solution, through action, experience and reflection, enabling more meaningful learning. In this sense, Levandowski and Schmidt (2010) state that there is a need to organise creative workshops, involving participants in a climate of trust and openness to dialogue, without prejudice, with diversified activities, both informative , to provide information and clarify doubts, and playful and experiential, to stimulate reflection on these themes.

FINAL CONSIDERATIONS

We found that sexuality is still seen as taboo, because for some adolescents it is accompanied by doubts, recriminations or traumas. The activities developed during the workshops showed the motivation and participation of the young people, making it possible to discuss the socio-cultural aspects of sexuality, as well as promoting reflection on various aspects involved in the subject.

We found that simple actions such as discussing sexuality at school, using methodologies that seek to promote dialogue and reflection on pre-established concepts, enable adolescents to develop and review attitudes, make value judgements and problematise issues such as the eroticisation of the image of women/men by the media, the correct use of condoms, teenage pregnancy and STDs, enabling adolescents to reflect on how to experience sexuality in a healthy and responsible way.

REFERENCES

BRAZIL. **National Curriculum Parameters:** third and fourth cycles: presentation of transversal themes. Brasília: MECSEF, 1998.

CANDAU, V. M. et al. **Oficinas pedagógicas de direitos humanos.** 2ª ed. Petrópolis, RJ: Vozes, 1995.

FIGUEIRÊDO, M. A. C. et al. Pedagogical workshop methodology: an extension experience with children and adolescents. **Revista Eletrônica Extensão Cidadã**. v.2, 2006.

LEVANDOWSKI, D. C.; SCHMIDT, M. M. Workshop on sexuality and dating for pre-adolescents. **Paidéia**, V. 20, N. 47, p. 431-436, 2010.

LIMA, E. B.; PAGAN, A. A. Sexuality, health and education: an overview of the school context. **Revista fórum identidades**, V. 8, N. 4, p. 90 - 109, 2010.

MOIZÉS, J. S.; BUENO, S. M. V. Understanding of sexuality and sex in schools according to primary school teachers. **Revista Escola de Enfermagem da USP**, V. 44, N. 1, p. 205 - 212, 2010.

Printed by Books on Demand GmbH, Norderstedt / Germany